What Shape Is It?

by Sheila Rivera

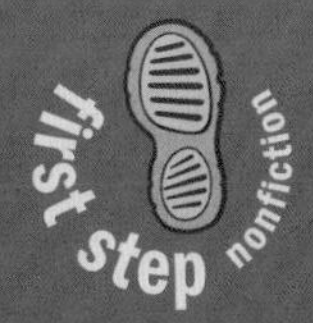

Lerner Publications Company · Minneapolis

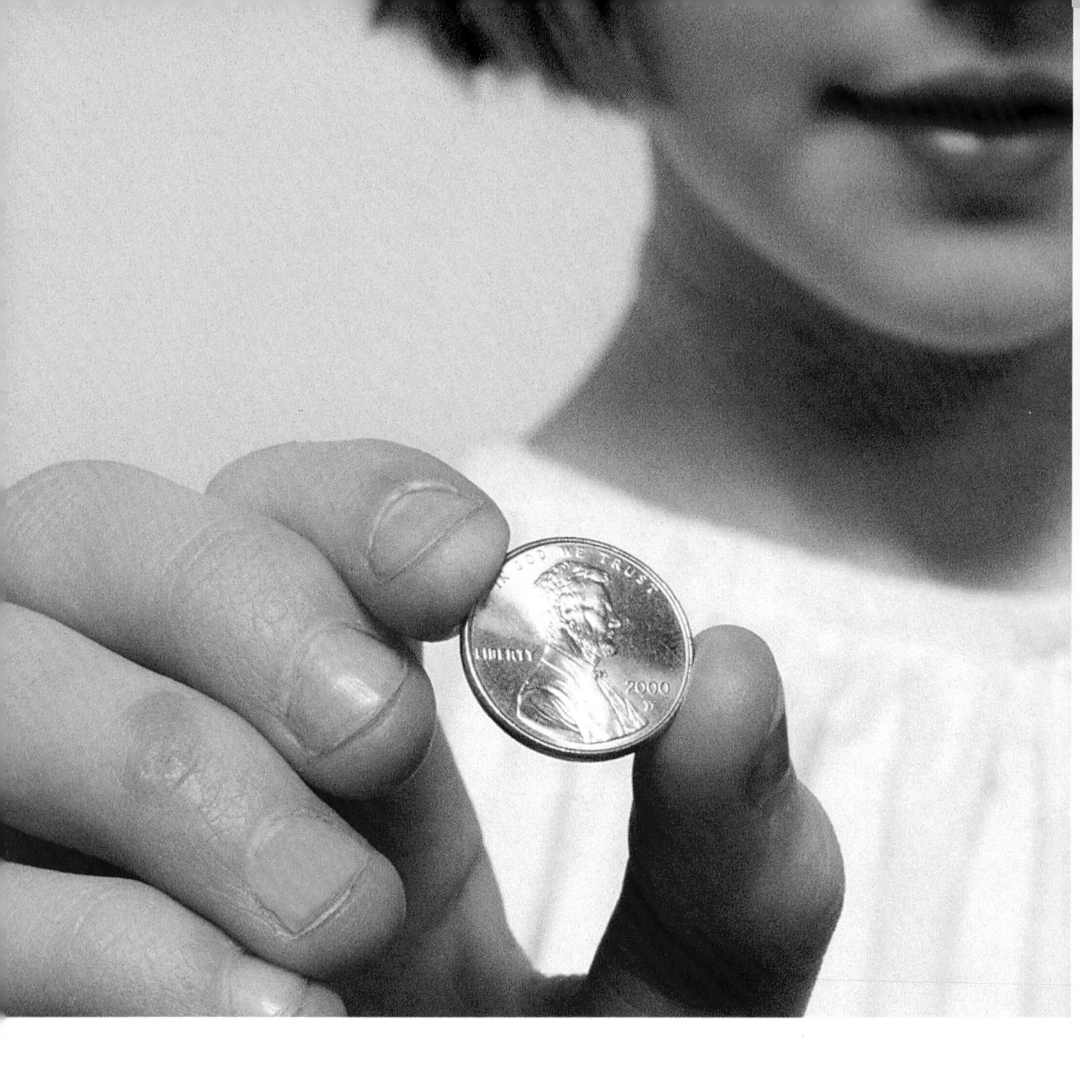

This penny is a circle.

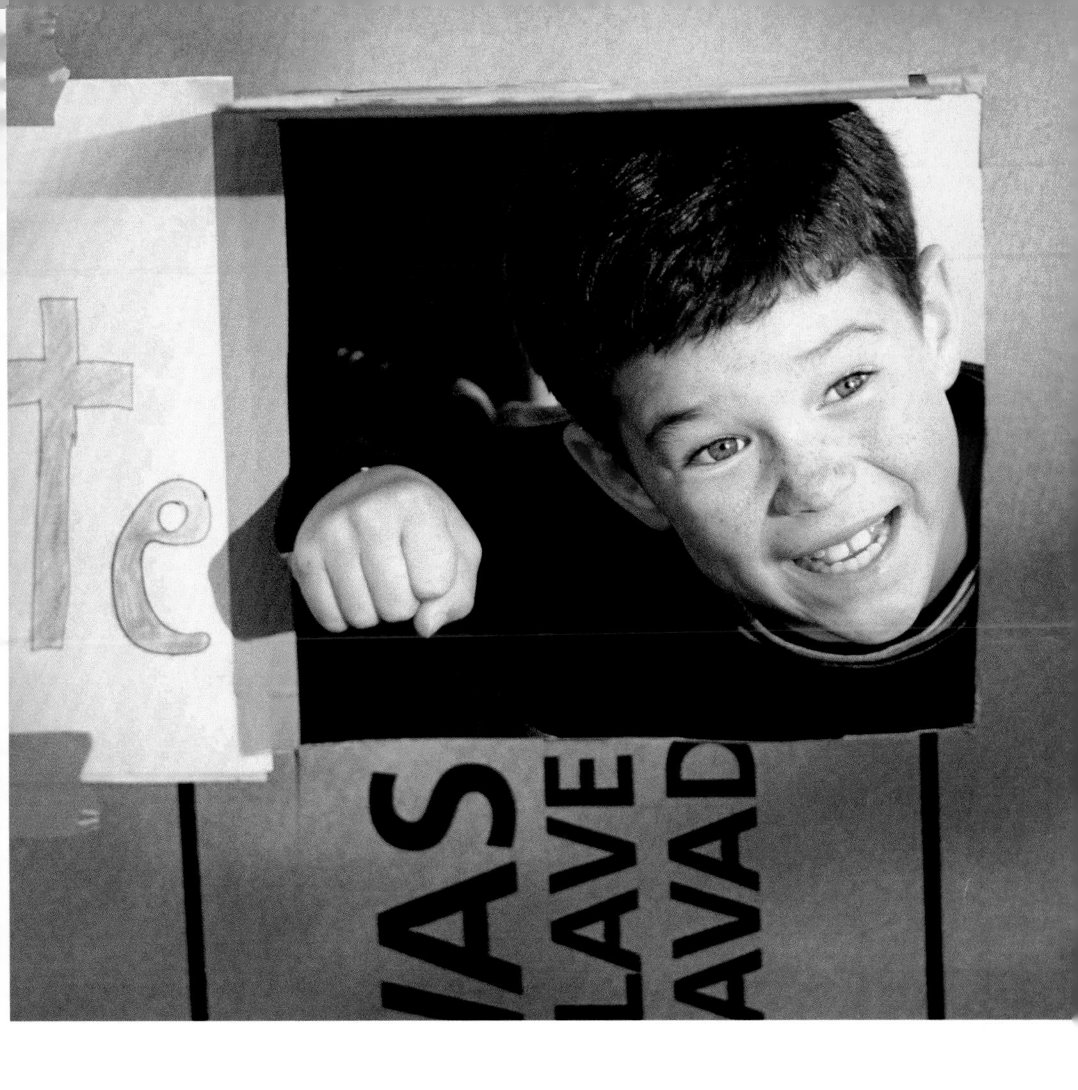

This window is a square.

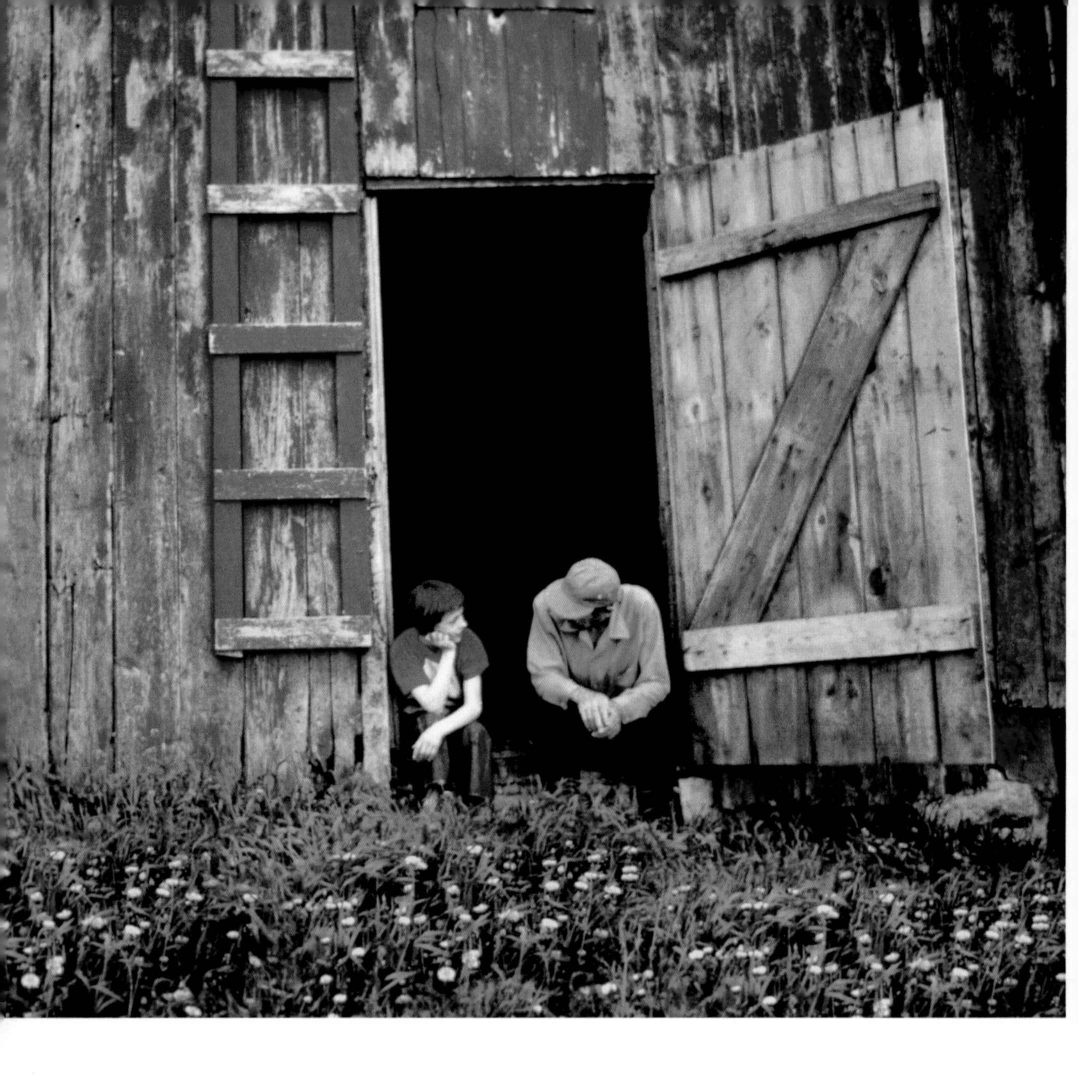

This door is a rectangle.

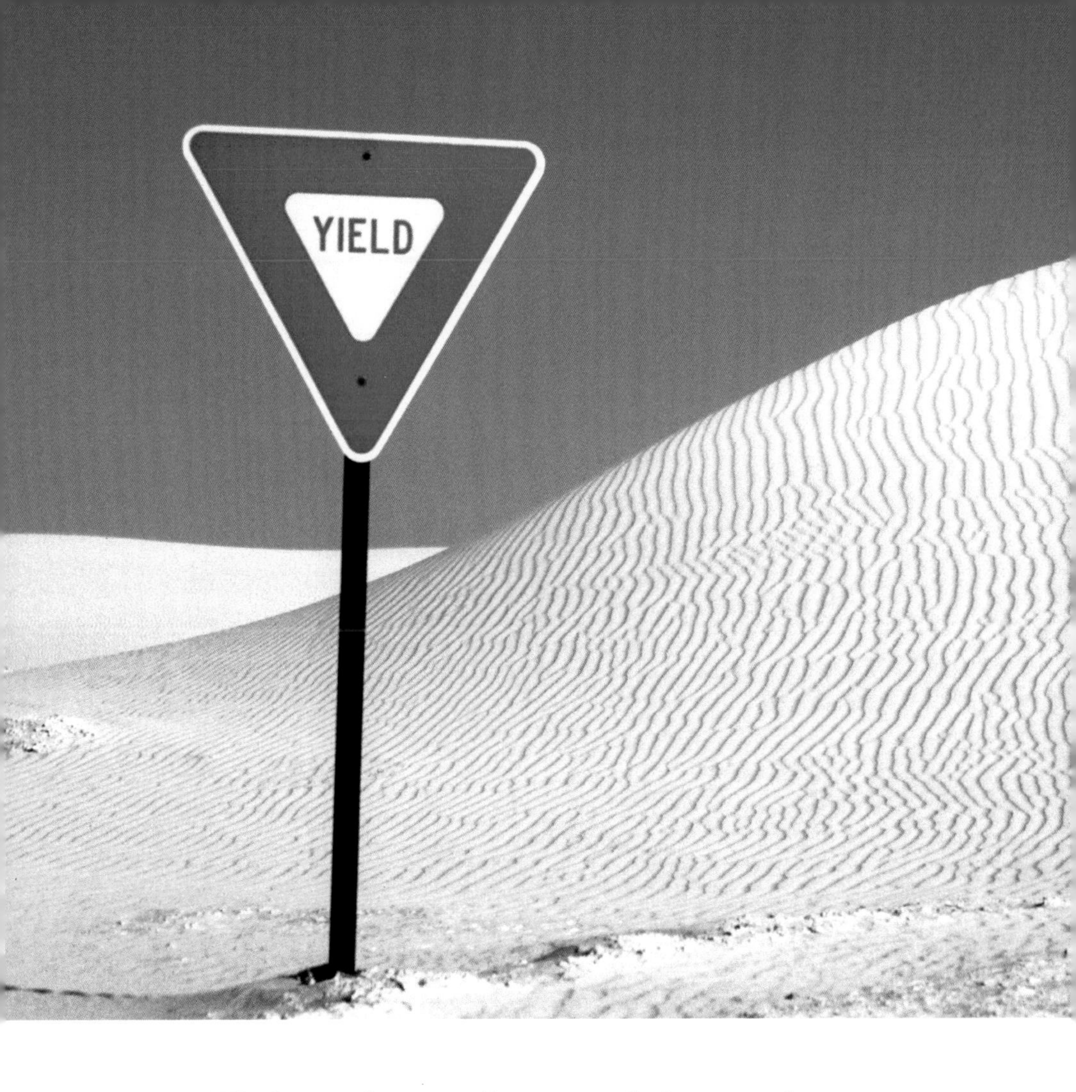

This sign is a triangle.

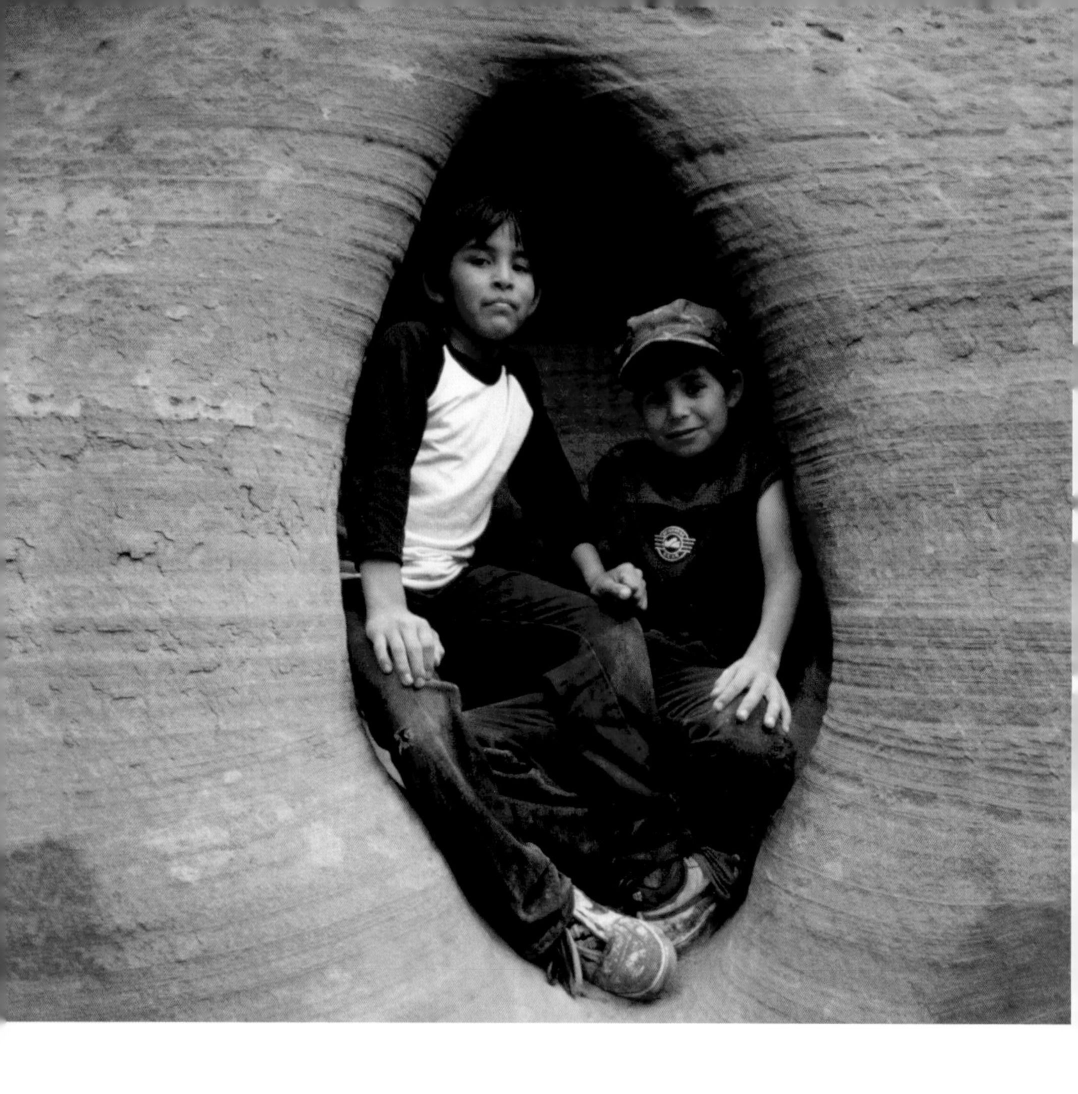

This hole is an oval.

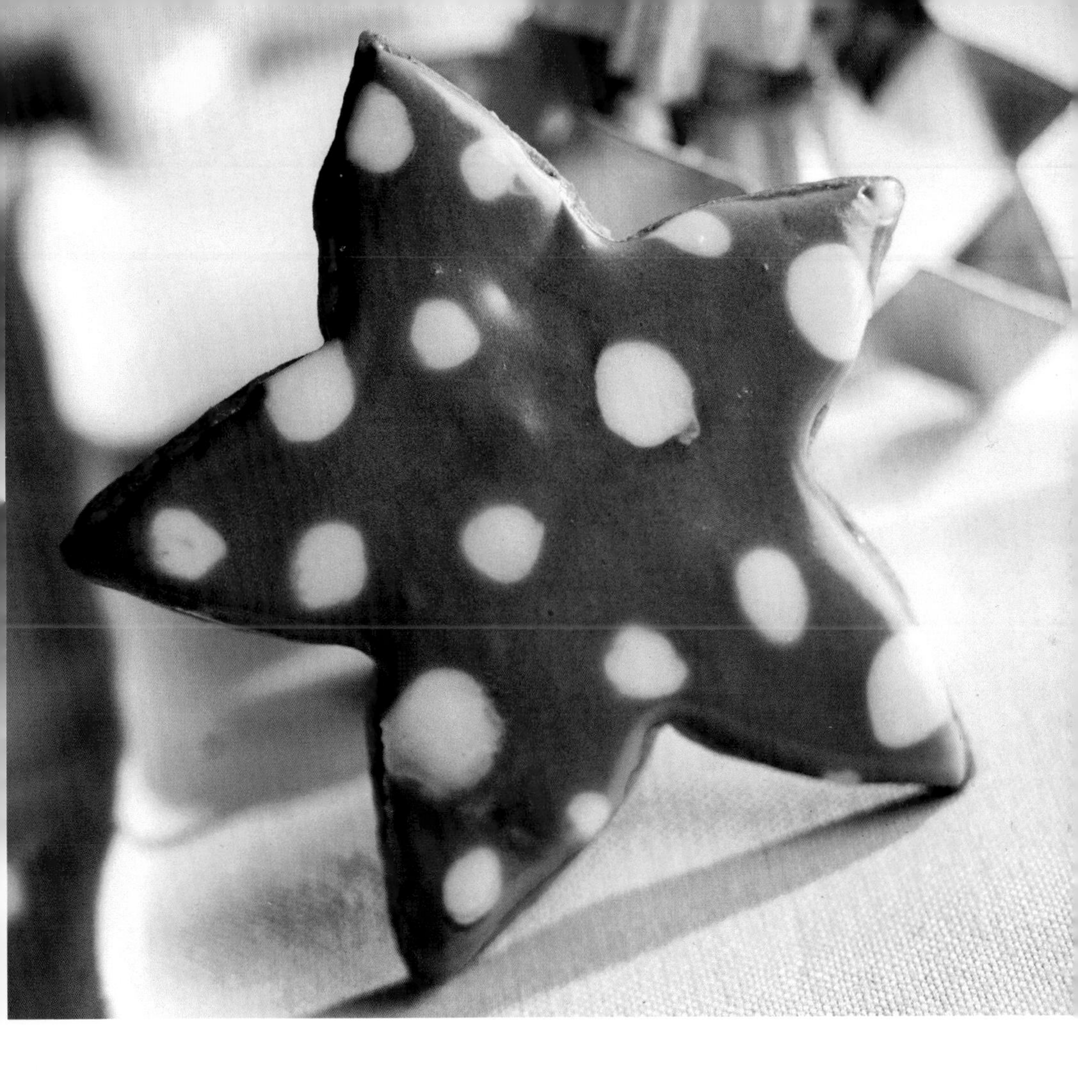

This cookie is a star.

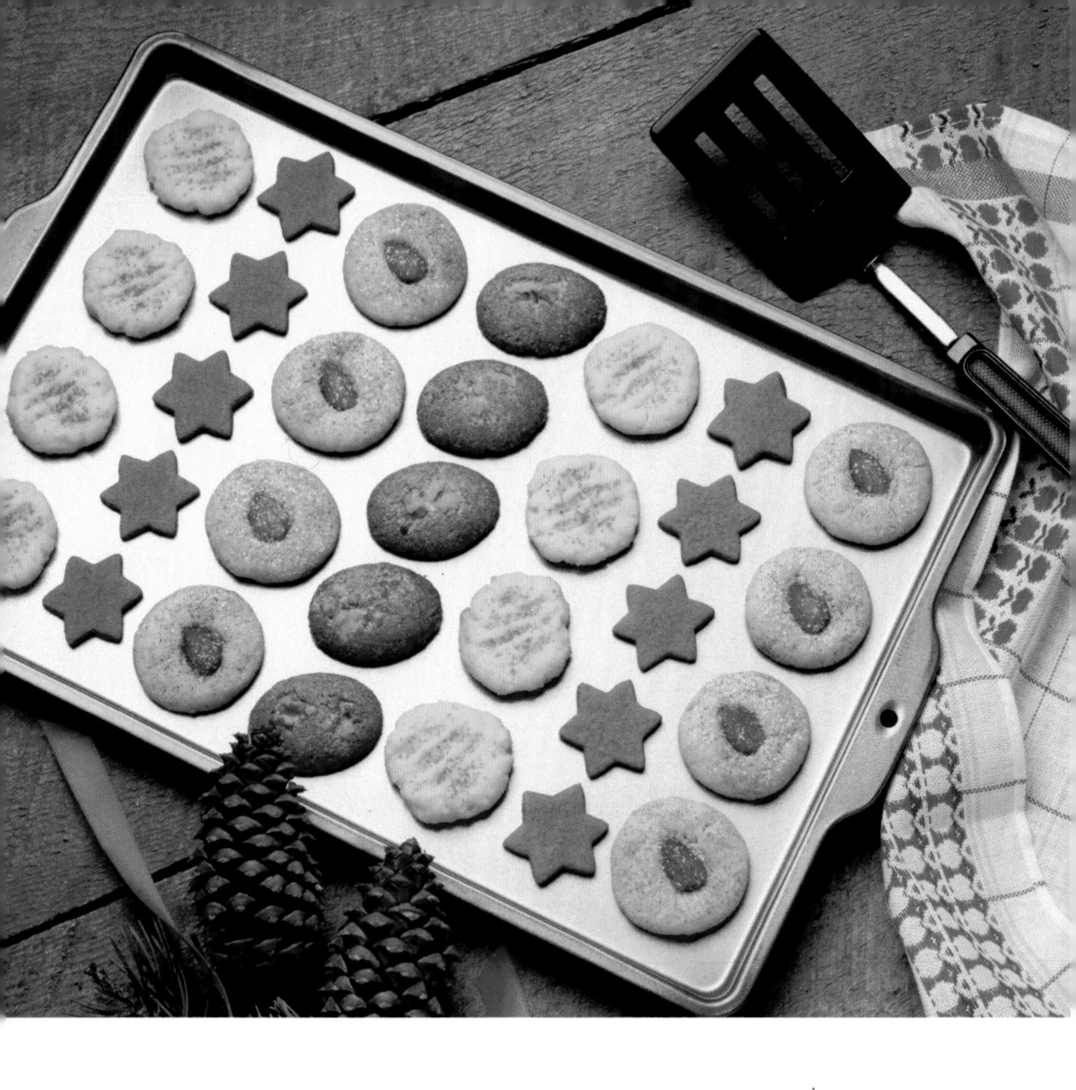

What shapes do you see?